Food
222

神奇泡泡

Bubble Magic

Gunter Pauli

[比] 冈特·鲍利 著

[哥伦] 凯瑟琳娜·巴赫 绘

颜莹莹 译

上海远东出版社

丛书编委会

主　任: 贾　峰

副主任: 何家振　闫世东　郑立明

委　员: 李原原　祝真旭　牛玲娟　梁雅丽　任泽林

　　　　王　岢　陈　卫　郑循如　吴建民　彭　勇

　　　　王梦雨　戴　虹　靳增江　孟　蝶　崔晓晓

特别感谢以下热心人士对童书工作的支持:

匡志强　方　芳　宋小华　解　东　厉　云　李　婧

刘　丹　熊彩虹　罗淑怡　旷　婉　杨　荣　刘学振

何圣霖　王必斗　潘林平　熊志强　廖清州　谭燕宁

王　征　白　纯　张林霞　寿颖慧　罗　佳　傅　俊

胡海朋　白永喆　韦小宏　李　杰　欧　亮

目录

Contents

一只座头鲸惊讶地发现，在海洋中央的一小片气泡上有一只蜗牛。他问道：

"你在这荒凉之地，头朝下做什么？"

"你太能兴风作浪了，拜托！别晃翻我的船！"蜗牛恳求道。

A humpback whale is surprised to come across a snail, on a tiny raft of bubbles, in the middle of the ocean. He asks:

"What are you doing out here in the middle of nowhere, and upside down, at that?"

"You are making too many waves, please! Don't rock my boat!" the snail pleads.

一只座头鲸惊讶地发现……

A hump whale is surprised ...

......这个气泡筏就是我的家。

... this bubble raft is my home.

"船？一堆泡沫算不上一条船。"

"这个气泡筏就是我的家。它是我用口水建造的。我把卵贮存在这儿，我的孩子们在这里玩耍，我还在这上面钓鱼。"

"这可真令人大开眼界！可是它看起来更像个气球，而不是一艘船。"

"Boat? A bunch of bubbles is hardly a boat."

"Well, this bubble raft is my home. I use my spit to build it. And it is where I store my eggs, and where my little ones play. It is also my fishing platform."

"Impressive! Even though it looks more like a balloon than a sea vessel."

"我一生都生活在水面上。我可以倒立着浮在水上，安静地吃着水母，不用为了一丁点口粮去跟海底那许多小鱼小虾们争抢。"

"好吧，泡泡蜗牛先生，我不吃水母，但是人们听到你吃水母会很高兴的。他们绝不喜欢游泳时身边有那么多的水母，真的是太多了。"

"I spend my life up here on the surface. Floating upside down, I can quietly get on with eating jellyfish, instead of fighting for every morsel on the sea floor where lots of other critters are looking for food."

"Well, Mr Bubble Snail, I don't eat jellyfish, but people will be very happy to hear that you do. Jellyfish are not exactly their favourite swimming partners, and now there are so many of them."

......吃水母。

... eating jellyfish.

你仍是最会捕鱼的。

you are still the best at fishing.

"吃你想吃的，做你想做的，这都随便。但事实是，如今以水母为食的<u>鲨鱼</u>和<u>金枪鱼</u>被捕捞得太多了，这导致水母大量繁殖。不过，你自己不就是个泡泡大师吗？"蜗牛问道。

"是的，我知道如何用气泡捕鱼。但跟你们不同，我们鲸不产卵，我们的孩子也没有一个安全的漂浮游乐场。"

"或许你的泡泡过一会儿就消失，但你仍是最会捕鱼的。还有谁能一下子就捕到一百万只磷虾呢？"

"It is up to you to eat and do what you like. But it is true that now so many of the sharks and tuna that feed on jellyfish are being caught, we see huge blooms of them. But aren't you a great master of bubbles yourself?" Snail asks.

"Yes, I have learnt how to fish using air bubbles. But unlike you, we whales don't have eggs. And we don't offer our little ones a safe, floating playground."

"The bubbles you make may disappear after a while, but you are still the best at fishing. Who else can catch a million krill in one go?"

"哦，我们很多动物都做过令人震惊的事。你见过蜘蛛夫人带着气泡下水吗？就像氧气罐一样。她一定是最早的潜水员。"

"真的吗？泡泡可太神奇了。"

"泡泡可能看起来很神奇，但其实它只是包裹在一层薄膜中的空气。"

"泡泡还有一个神奇之处，它们刚形成的时候有各种乱七八糟的形状，但很快都变成了圆形，像小球一样。"

"Oh, many of us do astounding things. Have you seen Mrs Spider carrying air bubbles, like an oxygen tank, when she goes in the water? She must have been the first diver."

"Really? Bubbles are magic."

"A bubble may seem like magic, but all it is, is air wrapped in a thin film."

"What else seems magical is that when they are first formed, bubbles have all kinds of crazy shapes, but in no time they all are round, like little balls."

就像氧气罐一样

like an oxygen tank

为什么它们是球形的？

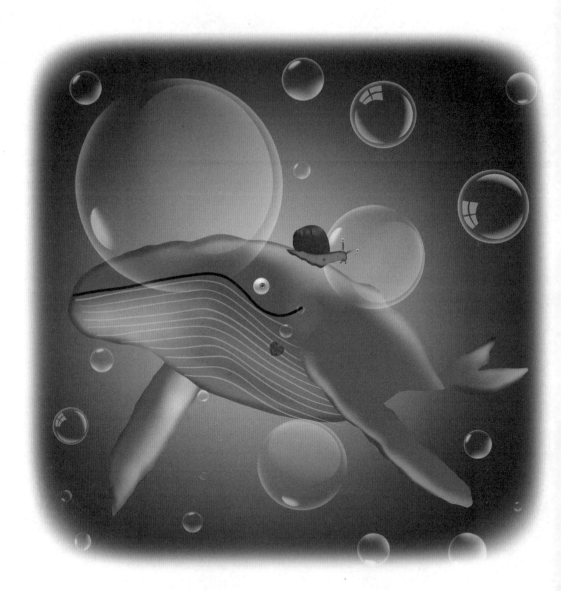

why they are spheres?

"哦，我喜欢观察气泡的形成。它们是如此的井然有序，不管是在水里还是在空气里，它们总是先变成球形然后再消失。"

"你能告诉我为什么它们是球形的吗？" 蜗牛问道。

"或者，我们应该问，为什么它们会破裂？"

"Oh, I do like to watch bubbles forming. They behave in such a disciplined way, creating order. In the water or in the air, they always turn into a sphere before they disappear."

"Would you mind telling me why they are spheres?" Snail asks.

"Or, should we ask the question, why do they pop?"

"你把我弄糊涂了，鲸夫人。我想知道为什么泡泡是圆的，而你却反问为什么它们会破裂。我们一次只谈一个问题，好吗？"

"我们为什么非要遵循从一个问题到下一个问题这样简单的逻辑呢？当我们的头脑自由地碰撞出火花，那一刻，不正是我们最有创意，而且我们的思想得到解放的时候吗？"

"Now you are confusing me, Mrs Whale. I wonder why bubbles are round and you respond by asking why they pop. Let's focus on one question at a time, shall we?"

"Why should we always follow a simple logic, going from one point to the next? Isn't it when we let our minds spark freely that we are most creative and innovative, and in that moment of chaos that our thinking is liberated?"

让我们一次只谈一个问题。

Let's focus.

用吸管沾水吹泡泡?

To pop a bubble with a wet straw?

"你又来了！用问题来回答问题……跟你说话可真费劲。我不知道你在说什么，那么……"

"……那么，你有没有听过泡泡在空气中破裂，或者从水里冒出来的声音？你见过它们变颜色吗？你试过用吸管沾水吹泡泡吗？我想知道为什么它不会破。我想知道为什么……"

"You are doing it again! Answering a question with a question… You are not easy to talk to, I'm afraid. And I have no idea what you are talking about now, so…"

" … So, have you ever listened to bubbles popping in the air, or bursting out of the water? Have you seen their colours change? Have you ever tried to pop a bubble with a wet straw? I wonder why it won't pop. I wonder why …"

"够了，谢谢！再这么下去，我会因绞尽脑汁去想这些神奇的泡泡而整夜失眠的。"

"是的，是时候弄清楚事情的原委了，这样我们才能搞定水母大量繁殖的问题。"

……这仅仅是开始！……

"Enough, please! As it is, I will have a sleepless night wracking my brain over the magic of bubbles."

Yes, it is time to learn how things work, so we can crack the problem of jellyfish blooms."

... AND IT HAS ONLY JUST BEGUN!...

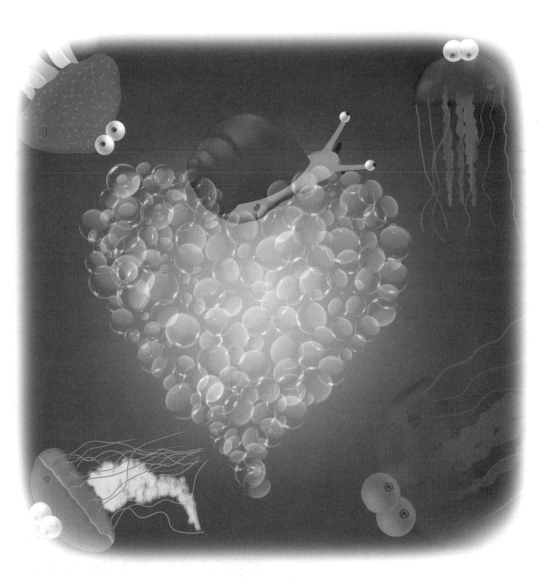

AND IT HAS ONLY JUST BEGUN! ...

Did You Know ?

你知道吗?

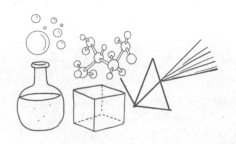

尽管气泡很脆弱，但却让我们能借此研究一些科学概念，如弹性、蒸发、温度、表面张力、化学反应、几何结构、光和颜色。许多物理学定律就是从简单的观察中发现的。

Bubbles, as fragile as they are, give us the opportunity to study scientific concepts such as elasticity, evaporation, temperature, surface tension, chemistry, geometry, light and colour. From simple observations we discover a great deal about the many laws of physics.

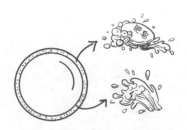

一个漂浮的泡泡就是在肥皂和水构成的薄膜内包裹着空气。肥皂泡的内外表面由肥皂分子构成，水分子构成了两层肥皂之间的薄层。

A floating bubble is air wrapped in a film made of soap and water. The inside and outside surfaces of a bubble consist of soap molecules. Water molecules form the thin layer between the two layers of soap.

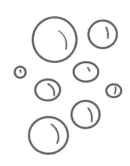

气泡总是会根据其内所含空气的体积收缩到尽可能小的形状。因为在体积固定的条件下球形的表面积最小，所以任何形状的气泡最后都会变成球形。

Bubbles always shrink to the smallest possible shape for the volume of air it contains. The search for efficiency transforms any bubble shape into a spherical one, which has the smallest surface area for the given amount of volume.

气泡破裂是因为两层薄膜之间的水蒸发了。空气越暖和，水蒸发得越快。这就是为什么在寒冷的冬天气泡可以持续几分钟，而在夏天仅仅几秒气泡就会破裂。

Bubbles pop because the water in between the two layers of film evaporates. The warmer the air, the faster the water will evaporate. That is why, on a cold winter's day, bubbles can last for several minutes, while in the summer they may pop in only a few seconds.

Cold atmospheric air makes bubbles fly higher. The warm air from our breath inside the bubble is lighter than the cold air surrounding the bubble. The greater the temperature difference between ambient air and body temperature, the higher the bubbles will fly.

冷空气使气泡飞得更高。我们呼出的热气在气泡内比其周围的冷空气更轻。环境空气与体温的温差越大，气泡飞得越高。

A bubble gets its colour from light waves reflecting from its inner and outer surfaces. As the water evaporates, the distance between the layers shrinks, and the colours change quickly.

气泡的颜色来自其内外表面反射的光波。随着水的蒸发，内外层之间的距离缩小，颜色迅速变化。

鲸在捕食磷虾和鲱鱼时会释放气泡。它们通过制造一圈上升的气泡来诱捕猎物。然后它们纵身潜入，把所有能吃到的东西一口吞下。

Whales release air bubbles when hunting for krill and herring. They trap their prey by creating a circle of rising air bubbles. Then they dive into this column, and eat all they can get in one single, very large, gulp.

海蜗牛从屈居海底与其他生物竞争抢食，进化为在海面的气泡上冲浪，成为僧帽水母这种有毒水母最大的捕食者之一。

The bubble snail evolved from dwelling on the sea floor, where it had to compete with other creatures for food, to surfing the surface of the ocean on a raft of bubbles where it is one of the largest predators of toxic jellyfish like the Portuguese man o' war.

Would you like to eat without others looking at what is on your plate?

你吃饭的时候希望没人注视盘里的东西吗？

The snail has to live upside-down to have food. Is this a fair compromise?

蜗牛吃东西时必须倒立。这是一个公平的妥协吗？

Do you want quick answers to single questions?

对于简单的问题，你想要得到快速的答复吗？

Are bubbles magic?

泡泡很神奇吗？

Do It Yourself!

自己动手!

Ask around to see who else enjoys playing with bubbles. Surely everyone does! But do people really know what a bubble is? Many will not aware of the fact that a bubble is formed by a layer of water between two layers of soap. Compile a list of interesting facts about bubbles and present this to your teacher. Sharing your list with your classmates will be an interesting way to introduce them to the wonders of physics. Your physics class will never be the same again!

问问周围有谁喜欢玩泡泡。当然每个人都喜欢! 但人们真的了解泡泡吗? 许多人并不知道其实泡泡是由两层肥皂之间的一层水膜形成的。罗列泡泡的趣事并交给老师。和你的同学们分享你的成果,这是向他们介绍物理奇迹的有趣方式。你的物理课再也不会和以前一样了!

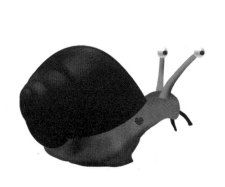

学科知识
Academic Knowledge

生物学	紫海蜗牛是全浮游蜗牛，是一种海洋腹足类动物；这种蜗牛一开始是雄性，后来变成雌性；卵在体内发育，后代一出生便能立即制造气泡；僧帽水母，一种海洋水螅类动物，有着致命的强大毒刺；僧帽水母是一种群体生物，它不是真正的水母；鲨鱼、金枪鱼和紫海蜗牛以水母为食；水母的触角可达60米长。
化 学	海蜗牛的气泡被裹在一层透明的几丁质层中；水母在高浓度的碳中分解，并以二氧化碳的形式释放出来；即使在水母死亡几天后，其刺丝囊仍然会蜇人；海水会使水母的刺细胞失去活性，而淡水会使它们恢复活性。
物 理	紫海蜗牛壳像纸一样薄，使其可以倒立漂浮；紫海蜗牛用它的斗篷（脚）捕捉空气，然后制造出气泡，并和它的黏液结合在一起；气泡不会飞，而是漂浮在密度稍高的空气中；浮力是向上的力；用肥皂减少表面张力；空气中的气泡总是会变成圆形，因为内外的作用力方向几乎相同；当大气泡破裂时，它们会形成更小的气泡；刺丝囊使水母的毒线伸展，像一个带弹簧的鱼叉。
工程学	球体是自然界最坚固、效率最高的形状；使用气泡捕鱼的新技术。
经济学	由于水母的大量繁殖，海滨发电厂和海水淡化厂承受了巨大的成本；水母通过船舶的压载水散布到世界各地；水母富含胶原蛋白，可以用来治疗关节炎，以及紧致皮肤减少皱纹。
伦理学	过度捕捞金枪鱼和鲨鱼减少了水母捕食者的数量，如果不采取措施保护它们，从浮游生物到鱼类的食物链将被破坏。
历 史	水母已经在海洋中生活了6亿年；林奈在1758年就对蜗牛进行过描述；它的名字来自希腊语；紫海蜗牛是海底蜗牛的后代；1991年，在"哥伦比亚号"航天飞机上，水母在零重力状态下进行了繁殖。
地 理	紫海蜗牛一生都在大海中度过；世界上各地的海洋里面都生活着水母。
数 学	肥皂泡以120度的等角三个一组连接在一起；运用几何分析最小表面积。
生活方式	在水母泛滥的海水中游泳；即时满足的文化，指对问题立即作答；我们在还没搞清运作机制的情况下就对事物加以应用，对其可能产生的后果也知之甚少。
社会学	社区渴望在混乱中创造秩序，一旦确立，便不愿改变或承担改变的风险。
心理学	因担心、恐惧或焦虑而失眠；由不确定、冲动和情绪引起的对即时满足的渴望（与延迟满足相对）。
系统论	由于全球变暖和过度捕捞，世界水母数量急剧增加，这可能加速海洋酸化；生态破坏的孪生诱因：全球变暖和海洋酸化；沿海地区受到塑料微粒和水母大量繁殖的影响；水母可以在缺氧和污染严重的海洋死亡区繁衍生息。

情感智慧
Emotional Intelligence

鲸

鲸惊讶地看着蜗牛倒立着浮在水面上。她不知道他在做什么，也不知道他的泡泡有何用处。当被问及她是如何使用泡泡时，鲸谦虚地回应，并指出其他生物也会使用气泡，如潜水蜘蛛。她简要地解释了气泡是如何产生的。鲸没有直接回答蜗牛提的问题，而是抛出了一连串的新问题，这让蜗牛感到不知所措。鲸知道探索未知才能获取新知识。她承认蜗牛控制着水母的大量繁殖，对生态系统作出了有益贡献。

蜗　牛

蜗牛担心鲸造成的海浪会摧毁他的家。他虽然弱小但不惧与鲸对抗。他了解导致环境问题的前因后果，也知道过度捕捞会导致鲨鱼和金枪鱼数量下降，而他能够有效控制水母数量。当鲸谦虚地介绍其他有趣生物时，蜗牛赞叹泡泡的神奇。他有勇气责备鲸，令其讲规矩、想答案。他赞赏鲸的智慧以及创造性的思维方式。他对水母繁殖的问题苦思冥想，担心又会有一个不眠之夜。

艺术
The Arts

让我们感受泡泡艺术的乐趣吧！你需要食物色素、洗洁精、水、一个浅盘子、一张纸和一根吸管。将水和肥皂液混合，插入吸管吹气，直到有一层厚厚的泡泡。加几滴食用色素。你可以使用多种颜色，但也不要混合太多，否则会变得浑浊。小心地在盘子上放一张纸，放在你的彩色泡沫上面。慢慢地把纸往下压，触碰到泡泡。气泡会粘在纸上。如果你把纸横着移动，你会得到一个大理石花纹效果。拿起纸，吹掉多余的颜料，然后平放晾干。你现在也是泡泡大师了！

思维拓展
Systems: Making the Connections

过去200年间，海洋吸收了人类排放的25%的二氧化碳。海洋对气候变化起着缓冲作用。随着海水中的二氧化碳不断增多，海洋酸性上升，这种平衡面临着挑战。海洋酸化已对紫海蜗牛等生物产生了腐蚀作用。对鱼类的过度捕捞正在影响生物多样性，破坏生态系统的平衡，这使食物链也深受影响。水母的总生物量已经是世界捕鱼总量的十倍之多。海洋污染——尤其是塑料微粒污染——使情况更加恶化。这些因素使海洋酸化加剧。水母的扩散繁殖会导致海洋"胶状化"。水母吞食大量的浮游生物，剥夺小鱼的食物，改变海洋食物链。海洋细菌能够吸收鱼类死亡后释放的碳、氮、磷和其他化学物质，但对水母却无能为力。这意味着水母的数量会不断增长。这种充斥着海洋的无脊椎动物的身体会分解成高碳生物质。细菌将其食用后转为二氧化碳呼出，增加了大气中的温室气体，并使海洋酸性升高。面对水母的过度繁殖，当前行之有效的办法不多。要解决这种杂乱的困境，关键在于清理海洋中的塑料垃圾，同时科学家和数学家应开始着手建立生态系统再生的模型。

动手能力
Capacity to Implement

自己动手做泡泡吧。其实很简单：只要把肥皂液和水按照1：6的比例混合就行了。确保你使用的是低过敏性肥皂，而不是那种由棕榈油制成的对生态有害的。最好使用蒸馏水。要想使泡泡持久，可以加入一勺玉米糖浆。下面将一根竹吸管插入容器，开始玩吹泡泡游戏吧。有一个有趣的实验，拿一把剪刀，将其浸入肥皂溶液中。接下来用你的吸管吹一个大泡泡，然后用剪刀的尖端戳入肥皂泡的内壁。肥皂泡将会自动包裹在任何湿的东西上，而不会破裂！

故事灵感来自

This Fable Is Inspired by

海伦·切尔斯基
Helen Czerski

　　海伦·切尔斯基在英国曼彻斯特附近长大。她在剑桥丘吉尔学院学习自然科学（物理）。2007年，她回到剑桥，致力于研究高速摄影，同时还潜心探索物理世界，并获博士学位。2007年至2008年，她加入美国加州圣地亚哥斯克里普斯海洋研究所，2010年至2011年在美国罗德岛海洋学研究所进行博士后研究工作。她对气泡和海洋进行了细致研究。之后她回到英国，先后在南安普顿大学和伦敦大学学院领导自己的研究项目。海伦专注于物理学和海洋。她开始在英国广播公司（BBC）工作，这使她有机会与广大观众分享她的科学热情。

图书在版编目（CIP）数据

冈特生态童书.第七辑：全36册：汉英对照 /
（比）冈特·鲍利著；（哥伦）凯瑟琳娜·巴赫绘；
何家振等译.—上海：上海远东出版社，2020
ISBN 978-7-5476-1671-0

Ⅰ.①冈… Ⅱ.①冈… ②凯… ③何… Ⅲ.①生态
环境–环境保护–儿童读物—汉英 Ⅳ.①X171.1–49

中国版本图书馆CIP数据核字（2020）第236911号

策　　划	张　蓉
责任编辑	程云琦
助理编辑	刘恩敏
封面设计	魏　来 李　廉

冈特生态童书

神奇泡泡

[比]冈特·鲍利　著
[哥伦]凯瑟琳娜·巴赫　绘

颜莹莹　译

记得要和身边的小朋友分享环保知识哦！
八喜冰淇淋祝你成为环保小使者！